LIFE OF STARS

From Origin to Oblivion

I0471569

ANANY AADIL

To all my friends and family, who were always supportive of me.

CONTENTS

INTRODUCTION

In this book, we will explore one of the most captivating subjects in all of science—stars. These glowing giants are more than just points of light in the night sky. They are engines of creation, harbingers of destruction, and silent witnesses to the story of our universe. From their ancient cultural importance to their powerful scientific role, this series of topics will guide you through the fascinating life cycle of stars and their impact on everything we know.

Humans have looked up at the stars for thousands of years, using them to tell stories, navigate oceans, and mark time. The first chapter explores how ancient civilizations understood stars and how their meanings evolved as science began to uncover the truth behind these brilliant objects.

Stars come in many forms—tiny red dwarfs to massive blue giants and more. In the second chapter, we'll look at the wide range of star types, what makes them different, and why those differences matter for their life cycle and behaviour.

Our own star, the Sun, is not just a source of light—it's a complex, layered system. In the third chapter, we'll break down the Sun's internal structure, from its blazing core to its outermost corona, and see how each part contributes to its immense power.

Stars are born in dark, cold clouds of gas and dust. With the help of gravity, pressure, and time, these clouds collapse to form new stars. The fourth chapter explains how stars come to life and the incredible processes that trigger their ignition.

Just like all living things, stars also die—but their deaths are far more dramatic. Whether it's a quiet fading or a massive explosion, the fifth chapter shows how a star's mass determines the way it will meet its end.

After death, a star leaves behind something strange and powerful—a white dwarf, neutron star, or even a black hole. In the sixth chapter, we'll examine these remnants and how they continue to influence the universe.

Where did the oxygen you breathe and the gold in your jewellery come from? Stars are cosmic factories, fusing lighter elements into heavier ones. The seventh chapter explores how stars build the elements that make up our world—and us.

Some stars don't just die—they destroy. From supernova bursts to silent killers, the eighth chapter highlights the most dangerous types of stars, their explosive behaviour, and what would happen if one turned its power toward Earth.

While the Sun can be fierce, it also protects us. With its solar wind and the bubble called the heliosphere, it shields Earth from deadly cosmic radiation. The final chapter reveals the Sun's hidden role as the guardian of our solar system.

HISTORICAL SIGNIFICANCE

For countless generations, humans have gazed up at the night sky, mesmerized by the vast number of twinkling lights scattered across the heavens. These tiny, distant dots have sparked curiosity and wonder for ages, leaving people to question, "What are these shining points of light that seem to watch over us?" At their core, stars are incredibly hot, dense masses that emit both light and heat. But this simple definition barely scratches the surface of what stars truly represent.

Significance in History:

Throughout history, stars have held a significant place in the cultural, religious, and scientific practices of civilizations around the globe. From their role in religious ceremonies and divination rituals to guiding navigators across vast oceans and helping societies organize time through calendars, stars have consistently been essential to human life. They were used to mark the changing of seasons, aiding agricultural planning, and became the backbone of ancient mythology, tying together stories of gods, heroes, and the cosmos.

Early astronomers made a key distinction between "fixed stars," which appeared to remain stationary on the celestial sphere, and "wandering stars," or planets, which moved against the backdrop of the fixed stars over days or weeks. Many ancient thinkers believed these stars were permanent fixtures attached to an eternal, unchanging heavenly sphere. This idea of an immutable cosmos shaped early astronomical models, with astronomers mapping the sky by grouping stars into recognizable patterns—asterisms and constellations. These patterns not only served as a visual aid but also helped track planetary motions and the position of the Sun. By observing the Sun's movement against the constellations, early societies were able to construct calendars, such as the Gregorian calendar still in use today, which is based on the Earth's tilt and its relation to the Sun.

Some of the earliest recorded attempts at mapping the stars date back millennia. In 1534 BC, ancient Egyptian astronomers created the oldest accurately dated star chart. Around the same time, Babylonian astronomers in Mesopotamia compiled star catalogues during the Kassite Period (1531 BC–1155 BC), laying the foundation for future astronomical work. By 300 BC, Greek astronomer *Aristillus*, along with *Timocharis*, produced the first Greek star catalogue. *Hipparchus*, another renowned Greek astronomer of the 2nd century BC, expanded this knowledge by cataloguing over 1,000 stars. His detailed observations even led to the discovery of a "nova" or new star, further demonstrating that the heavens were not as immutable as once thought. Many of the constellations and star names we use today are derived from Greek astronomy, thanks to the groundwork laid by scholars like *Hipparchus* and *Ptolemy*.

Despite the belief that the heavens were unchanging, astronomers from various cultures observed phenomena that contradicted this idea. For instance, Chinese astronomers were among the first to record supernovae. In 185 AD, they documented SN 185, the first recorded supernova. The brightest stellar event in human history occurred in 1006 AD with the supernova SN 1006, observed by Egyptian astronomer *Ali ibn Ridwan* and Chinese astronomers. The Crab Nebula, born from the SN 1054 supernova, was also witnessed by Chinese and Islamic astronomers, furthering the understanding that stars could change and even end their lives in spectacular explosions.

Medieval Islamic astronomers made notable contributions to star mapping and stellar observations. They developed sophisticated astronomical instruments to calculate the positions of stars, and their observations formed the basis of many modern star catalogues. *Abd al-Rahman al-Sufi's* **Book of Fixed Stars**

(964 AD) catalogued numerous stars, star clusters, and even galaxies like Andromeda. Persian polymath *Abu Rayhan Biruni*, writing in the 11th century, was one of the first to describe the Milky Way as a cluster of nebulous stars. Meanwhile, Andalusian astronomer *Ibn Bajjah* hypothesized that the Milky Way was composed of closely packed stars, suggesting that their light appeared continuous due to refraction from sublunary material. These early observations paved the way for future studies on the nature and composition of galaxies.

In Europe, early modern astronomers began to challenge the long-held notion of the immutability of the heavens. *Tycho Brahe*, for instance, recorded the appearance of new stars, or novae, further questioning the idea of a fixed celestial sphere. Giordano Bruno, in 1584, advanced the revolutionary idea that stars were similar to the Sun and might have planets orbiting them. This notion, though radical at the time, was built on earlier ideas from ancient Greek philosophers like *Democritus* and *Epicurus*, as well as Islamic scholars such as *Fakhr al-Din al-Razi*. By the 17th century, this concept of stars being akin to the Sun gained wider acceptance, and Isaac Newton suggested that stars were evenly distributed across the cosmos to explain the lack of a gravitational pull from these distant bodies, an idea influenced by theologian Richard Bentley.

By the 18th century, the study of stars had taken on a new level of precision. In 1667, Italian astronomer *Geminiano Montanari* was the first to record variations in the brightness of the star *Algol*. Edmond Halley, famous for his work on comets, made groundbreaking measurements of the proper motion of nearby stars, revealing that even "fixed" stars shifted position over time. William Herschel, one of the most influential astronomers of the late 18th century, sought to map the distribution of stars in the sky. His extensive star counts led him to conclude that stars were concentrated toward the centre of the Milky Way galaxy. His son, John Herschel, continued this work in the southern hemisphere, confirming his father's findings. William Herschel also discovered that some stars existed in pairs, forming binary star systems.

In the 19th century, advances in technology and observational methods furthered the scientific study of stars. Joseph von Fraunhofer and Angelo Secchi pioneered the field of stellar spectroscopy, using absorption lines in stellar spectra to distinguish stars based on their chemical compositions and temperatures. Secchi's classification of stars into spectral types laid the groundwork for the modern stellar classification system, which was refined by Annie J. Cannon in the early 20th century.

Friedrich Bessel, in 1838, made the first direct measurement of the distance to a star using the parallax method, demonstrating the vast distances separating stars in the cosmos. By observing binary star systems, astronomers were able to calculate the masses of stars, while Felix Savary's work on orbital elements in 1827 marked a key milestone in the study of these systems.

The 20th century saw rapid progress in stellar astronomy, fuelled by advances in technology. Karl Schwarzschild's discovery that a star's colour could be used to estimate its temperature revolutionized stellar classification. The photoelectric photometer allowed astronomers to precisely measure a star's brightness across different wavelengths, and Albert A. Michelson's use of an interferometer in 1921 enabled the first measurements of a star's diameter.

During this period, astronomers also developed models explaining the internal structure and life cycles of stars. The Hertzsprung-Russell diagram, created in 1913, became a cornerstone of stellar astrophysics, while Cecilia Payne-Gaposchkin's groundbreaking 1925 PhD thesis established that stars were composed primarily of hydrogen and helium. Advances in quantum physics further deepened our understanding of stellar spectra, enabling scientists to determine the chemical compositions of stars.

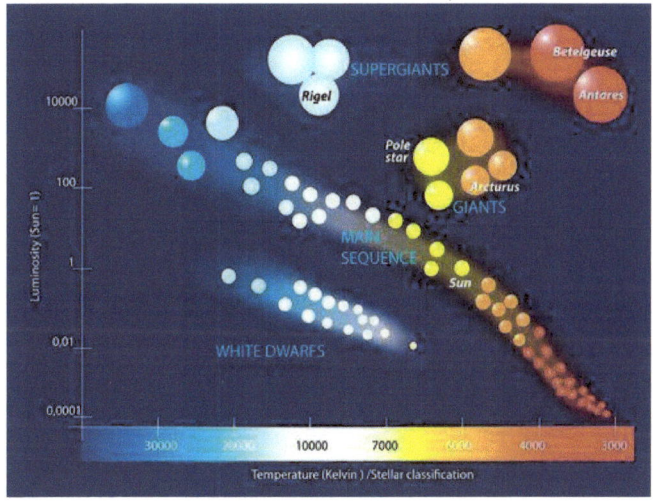

Fig. 1 The Hertzsprung-Russell Diagram

While individual stars are mostly observed within our galaxy, modern astronomy has extended our view to stars in other galaxies, including those in the Virgo Cluster. With the help of gravitational lensing, astronomers have even observed a star, named *Icarus*, located 9 billion light-years away, pushing the boundaries of what we know about the universe and the stars that populate it.

TYPES OF STARS

Thereare a number of types of stars found in the universe, the following is the list of the most significant ones:

Blue Supergiants:

These are among the most massive and luminous stars ever discovered in the universe. These colossal stellar giants are notable for their immense size, intense brightness, and extremely high surface temperatures. Theoretical models suggest that some of the very first stars formed after the Big Bang, known as *Population -III* stars,

Fig. 2 A Blue Supergiant

were blue supergiants. These ancient stars are believed to have been crucial in shaping the early universe, although none have been observed directly yet due to their short lifespans and unstable nature.

Blue supergiants are characterized by their remarkable instability. Unlike smaller stars, which can remain relatively steady for billions of years, these massive stars burn through their nuclear fuel at an astonishing rate, causing frequent fluctuations in their brightness and size. Their luminosity is classified as **Class I** (Supergiant Class), the highest possible classification, indicating that they emit extraordinary amounts of energy. This high energy output is a direct result of their incredibly hot surface temperatures, which range from about 9,000°C to as high as 40,000°C. Such intense heat makes them glow with a brilliant blue hue, which gives them their name.

These stars are not only significant in our understanding of stellar evolution but also play a key role in the broader processes of cosmic evolution. As they exhaust their nuclear fuel, blue supergiants are prone to catastrophic end-of-life events, such as supernovae, where they explode and release vast amounts of heavy elements into space, contributing to the formation of new stars and planets.

One of the most famous and well-studied examples of a blue supergiant is *Rigel*, the brightest star in the constellation of Orion. This stellar giant stands out due to its incredible mass and brightness. With a mass approximately 20 times greater than that of our Sun, Rigel is a colossal force in the universe. Its luminosity, or the amount of light and energy it emits, is even more astonishing—Rigel shines with a brightness about 117,000 times greater than the Sun, making it one of the most luminous stars visible from Earth.

Rigel's immense energy output and size place it among the most powerful stars in the night sky.

Red Supergiants:

Red supergiants rank among the largest stars known to exist in the universe, defined not by their mass or brightness but by their sheer volume. Despite their immense size, they are not necessarily the most massive or luminous stars, particularly when compared to other stellar giants like blue supergiants. Classified under the supergiant luminosity class, these stars are distinct for their cooler surface temperatures and enormous physical dimensions.

Unlike the intensely hot blue supergiants, red supergiants are cooler and much larger in size. Their surface temperatures are typically around 4,500°C, much lower than their blue counterparts. Interestingly, many stars that begin as blue supergiants or even smaller stars may expand dramatically in their later life stages, evolving into red

Fig. 3 Betelgeuse - A Red Supergiant

supergiants as they burn through their nuclear fuel. This process leads to a dramatic increase in their size, even as their outer layers cool. These fascinating transformations and the immense size of red supergiants make them a crucial area of study in understanding stellar evolution and the life cycle of massive stars.

Red supergiants are relatively rare in the universe, yet their immense size and brightness allow them to be seen from great distances. Many of these stars are also variable, meaning their brightness fluctuates over time, making them prominent and well-known examples of naked-eye stars. Some of the most famous red supergiants include:

- *Antares A* in the constellation Scorpius.
- *Betelgeuse*, a bright star in Orion
- *Epsilon Pegasi* in Pegasus.
- *Zeta Cephei* from the constellation *Cepheus.*
- *Lambda Velorum* located in *Vela.*
- *Eta Persei* in Perseus.
- *31 and 32 Cygni*, a pair in *Cygnus.*
- *Psi Aurigae* in the *Auriga* constellation.

- *119 Tauri*, another bright red supergiant in Taurus.

These stars are some of the best-known examples of red supergiants that can be observed without the need for telescopes. Their variability and distinctive red hue make them stand out in the night sky.

Main sequence stars:

Often referred to as dwarf stars, are considerably smaller than their larger stellar counterparts, yet they are by far the most abundant type of star in the universe. These stars are the backbone of the cosmic landscape, with the Sun being the most well-known example. Despite their relatively modest size, main sequence stars play a vital role in sustaining long-term stability, making them ideal environments for the potential development of life.

These stars are known for their remarkable stability, as they spend the majority of their lives in a balanced state, where the outward pressure from nuclear fusion in their cores is counteracted by the gravitational force pulling inward. This equilibrium allows them to shine steadily for billions of years. Main sequence stars typically have surface temperatures ranging from 5,000°C to 6,000°C, while their cores can reach an astonishing temperature of over 14 Million°C.

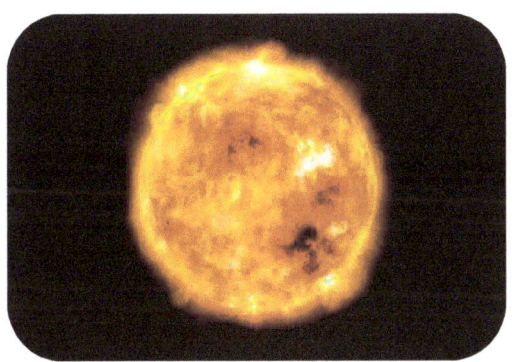

Fig. 4 Sun - A Main sequence star

Their stability and prevalence throughout the universe make main sequence stars a fundamental element in the evolution of galaxies and the potential for life-bearing planets.

Red-dwarf:

A red dwarf is the smallest type of star found on the main sequence of stellar evolution, and they initially formed as yellow main sequence stars. However, due to their relatively low mass, these stars were unable to sustain the necessary energy output to remain yellow, gradually cooling and turning red over time. Red dwarfs are by far the most abundant stars in the Milky Way, especially in the vicinity of the Sun.

Despite their abundance, individual red dwarfs are difficult to observe because of their faint luminosity. In fact, no red dwarfs that meet the stricter

Fig. 5 A Red Dwarf

definitions are visible to the naked eye from Earth. One notable example being *Proxima Centauri*, the closest star to the Sun, is an example of a red dwarf, and fifty out of the sixty nearest stars are also of this type. It is estimated that about three-quarters of all stars undergoing nuclear fusion in the Milky Way are red dwarfs.

Brown Dwarf:

Brown dwarfs are substellar objects with a mass greater than the largest gas giant planets but less than the smallest main-sequence stars. These objects are often considered "failed stars" because they do not have enough mass to sustain nuclear fusion, the process that powers true stars. Due to their relatively low luminosity, brown dwarfs are difficult to detect and observe. While they are not massive enough to be classified as stars, they still possess a significant gravitational pull.

Fig. 6 Illustration of a Brown Dwarf

Brown dwarfs occupy a unique space between planets and stars, often referred to as the middle ground between the two. Oftentimes casually named "Big Jupiters," they resemble gas giants like Jupiter but are many times more massive. Many planetary scientists believe there could be a significant number of brown dwarfs scattered throughout the universe. At one point, a theory suggested that the Sun might have a brown dwarf as a binary companion, called "*Nemesis*." However, this theory has been largely discredited due to a lack of supporting evidence.

Hypergiants:

The Hypergiant star is the largest and most luminous type of star in the universe, once referred to as a "Super-Supergiant" or "Extreme-Supergiant" before the term Hypergiant was adopted. These stars surpass even red supergiants in size, though they are incredibly rare. Hypergiants are distinguished by their immense luminosity, mass, and size, as well as their significant mass loss due to intense stellar winds. They are classified under luminosity **Class Ia-0** Their extreme brightness and powerful winds make them standout celestial objects, but they are unstable and lose mass rapidly.

Yellow hypergiants are thought to be evolved red supergiants that have shed much of their outer atmosphere and hydrogen, marking an advanced stage in their stellar life cycle. The largest known star is a Hypergiant named *UY Scuti*. Other examples include *The Pistol Star*, *Rho Cassiopeiae*, *Mu Cephie* and the previous record holder for the largest known star *VY Canis Majoris*.

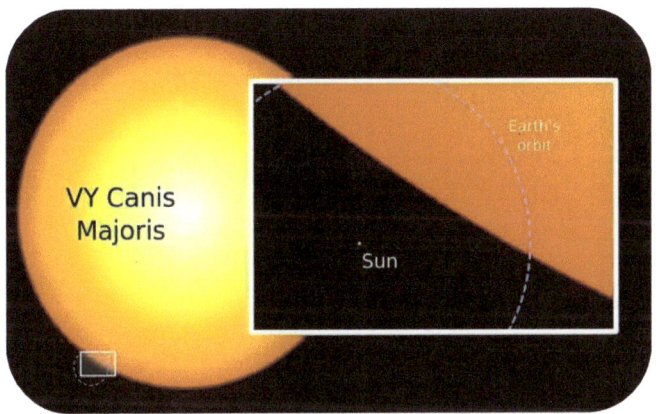

Fig. 7 Comparison of VY Canis Majoris against Earth's Orbit

THE STRUCTURE
OF THE SUN

The Sun, just like Earth, is formed of multiple layers that govern the properties of this celestial body.

Photosphere:

The part of sun that is directly observable with visible light. From which the layer below becomes opaque.

Photons produced in this layer escape the Sun through the transparent solar atmosphere above it and become solar radiation aka sunlight. The photosphere is tens to hundreds of kilometres thick, and is slightly less opaque than air on Earth. Because the upper part of the photosphere is cooler than the lower part, an image of the Sun appears brighter in the centre than on the edge or limb of the solar disk, in a phenomenon known as limb darkening.

Since the sun is a gaseous celestial body, it does not have a clearly defined surface. Its visible parts are usually divided into an "Atmosphere" consisting a "Photosphere".

Atmosphere:

The Sun's atmosphere is made up of five distinct layers: the photosphere, the chromosphere, the transition region, the corona, and the heliosphere.

The coolest layer of the Sun is a temperature minimum region that extends to about 500 km above the photosphere, where the temperature is approximately 3,800°C. This part of the Sun is cool enough to allow for the presence of simple molecules like carbon monoxide and water, which can be detected through their absorption spectra. In contrast, the chromosphere, transition region, and corona are significantly hotter than the Sun's surface.

Just above the temperature minimum region is the chromosphere, a layer about 2,000 km thick that is characterized by a mix of emission and absorption lines. Its name, derived from the Greek word chroma meaning colour, refers to its visibility as a flash of colour during the beginning and end of total solar eclipses. The temperature within the chromosphere gradually rises with altitude, reaching up to about 20,000°C near the top.

Above the chromosphere lies the transition region, a very thin layer—only about 200 km thick—where temperatures increase sharply from around 20,000°C to coronal temperatures close to 1 Million°C. This rapid temperature rise is due to the full ionization of helium in the transition region, which drastically reduces the cooling of the plasma. The transition region is not fixed at a specific altitude and instead forms a fluctuating boundary around features like spicules and filaments, constantly shifting in chaotic motion. It is difficult to observe this region from Earth's surface, but it is easily detected by instruments in space that can sense extreme ultraviolet radiation.

The next layer is the corona, the Sun's outer atmosphere. The lower corona lies close to the Sun's surface. The average temperature of the corona and solar wind ranges from about 1 - 2 Million °C, but in its hottest areas, temperatures can reach between 8 Million°C to 20 Million°C. Although there is no comprehensive theory yet to fully explain the corona's extreme heat, magnetic reconnection is known to contribute to some of this energy. The corona is vastly larger than the Sun's photosphere and makes up the Sun's extended atmosphere, stretching far beyond the Sun itself.

The heliosphere, which represents the Sun's outermost and most diffuse atmospheric layer, is filled with a stream of charged particles known as solar wind plasma. This solar wind flows outward from the Sun in all directions, extending throughout the heliosphere. As it moves, the solar wind causes the Sun's magnetic field to stretch and twist, forming a spiral shape due to the Sun's rotation. This vast region acts as a protective bubble, shielding our solar system from high-energy cosmic radiation.

Convective Zone:

Between approximately 70% of the Sun's radius and just below its visible surface, the solar plasma becomes neither hot nor dense enough to allow radiation to efficiently carry energy outward. This causes convection to take over as the dominant method of heat transfer. In this convection zone, the solar material moves in an orderly pattern, developing into thermal cells that carry heat toward the surface. These convective movements are somewhat similar to the weather cells found in Earth's atmosphere, where warm air rises and cooler air sinks, creating circulation patterns. This process is crucial for transporting energy from the Sun's interior to its surface.

Tachocline:

The radiative zone and the convective zone of the Sun are divided by a transitional layer known as the tachocline. This is where the dynamics of the Sun shift drastically, as the uniform rotation of the radiative zone transitions into the differential rotation of the convective zone. In this region, the difference in rotational speeds creates a significant shear effect, where horizontal layers of solar material slide past each other. It is currently theorized that within the tachocline, a magnetic or solar dynamo operates, generating the Sun's powerful magnetic field through this shear-driven movement.

Radiative Zone:

Convection does not begin until much closer to the Sun's surface. As a result, between approximately 20-25% of the Sun's radius and extending outward to about 70% of its radius lies the radiative zone. In this region, energy is transferred primarily through thermal radiation rather than by convection. Here, photons gradually carry energy outward from the Sun's core, scattering repeatedly as they travel, rather than through the large-scale movement of solar material, which dominates in the outer layers.

Core:

The core is the central 20-25% of the Sun's radius, where the temperature and pressure are extremely high, creating the conditions necessary for nuclear fusion to take place. During fusion, hydrogen nuclei combine to form helium, releasing vast amounts of energy in the process. Over time, the core gradually accumulates more helium as a result of this fusion. This region is the primary source of the Sun's energy, producing nearly all of its thermal output. In fact, 99% of the Sun's power is generated within the innermost 24% of its radius, with fusion activity almost non-existent beyond 30%. The energy produced in the core is transferred outward through the Sun's layers, heating them as it moves, eventually reaching the photosphere, from where it radiates into space.

FORMATION OF
STARS

Stars are the colossal powerhouses of the universe, responsible for both creation and destruction. In the previous topic, we explored the different types of stars, but how exactly are they born?

Everything in the universe has an origin. There was once a time, about 13.8 billion years ago, when not a single star existed. Back then, right after the Big Bang, the universe was filled primarily with hydrogen gas, the simplest and most abundant element. It was from this primordial hydrogen that the first stars eventually emerged. These early stars, known as Population III stars, were the largest and most luminous, shining a million times brighter than the Sun.

But let's not dive too deeply into these early giants just yet. For now, we'll focus on the process of how stars, in general, are formed.

Stars are born within vast interstellar clouds, known as nebulae, which consist of cold molecular gases and particles moving at different speeds. Gravity gradually pulls these particles together, causing them to collide. As they do, kinetic energy is released as heat and light. Over time, these particles merge, growing larger and strengthening their gravitational pull, which attracts even more material from the surrounding nebula. Eventually, the gravitational force becomes too strong for the gas and dust to resist, leading the core of the cloud to collapse and form a protostar.

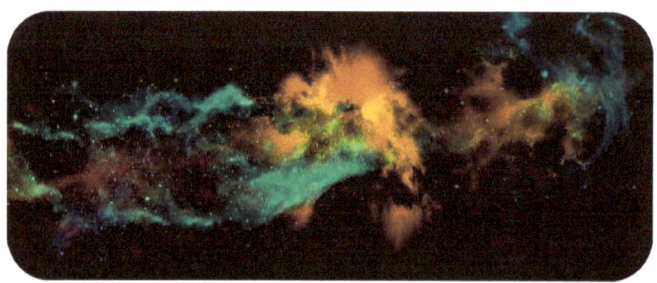

Fig. 8 A Nebula - Cloud of Gas and Dust

Star formation occurs in dense regions of these molecular clouds, often called stellar nurseries. As the cloud collapses, it fragments into smaller pieces, which are destined to become stars. These fragments continue to shrink and condense due to gravity. As they do, gravitational potential energy is converted into heat, raising the temperature and preventing the fragments from breaking down any further. The fragments turn into rotating spheres of gas, acting as stellar embryos.

The collapsing protostellar cloud continues to shrink as long as the energy generated by gravitational compression can escape. However, once the cloud becomes opaque to its own radiation, it can no longer radiate away this energy, and other processes must kick in to get rid of the excess energy. As the cloud's density increases, particularly in the central region, the inner part becomes opaque first. The protostar keeps growing as material accretes onto it, often through the formation of a circumstellar disk.

When the core becomes hot and dense enough, deuterium fusion begins, producing outward radiation pressure that slows the collapse. Eventually, hydrogen fusion ignites in the core, signalling the birth of a main-sequence star, as the leftover material surrounding the star is blown away.

Stars are the primary architects of the universe. Without them, the cosmos would largely consist of just hydrogen gas. These nuclear powerhouses not only create energy but also form the building blocks of everything we see.

As mentioned before, the extreme pressure and gravitational forces in the cores of stars cause the nuclear fusion of hydrogen atoms into helium. This process continues, eventually forming heavier elements like carbon and oxygen, which are essential for the creation of planets, life, and everything around us.

The Sun, as the most notable star, plays a crucial role in supporting life on Earth. Its energy fuels the planet's ecosystems, making it the main source of life-sustaining energy.

Thus, we can say that stars are far more important entities in the universe than merely twinkling in the night sky. They play a fundamental role in shaping the cosmos, creating essential elements, and driving the processes that allow life to exist. Their influence extends well beyond their visible glow, making them pivotal in the ongoing evolution of the universe.

DEATH OF STARS

I have already discussed several types of stars and explained how they are born in the vast clouds of nebulae.

However, just as life is a natural part of the universe, so is death. This applies not only to living organisms but also to stars. Stars are not eternal—they have life cycles just like everything else in the cosmos. After shining for millions or even billions of years, stars eventually reach the end of their lifespans and die in different ways depending on their mass.

One of the simplest ways to estimate a star's lifespan is by considering its size. There is a general rule in astrophysics: the larger the star, the shorter its life. This might seem surprising at first, but it makes sense when you understand how stars work. Bigger stars have much more mass, which leads to higher pressure and temperature in their cores. This causes them to burn through their hydrogen fuel at a much faster rate than smaller stars. So, even though a massive star contains more fuel, it uses it up quickly, leading to a shorter life.

In contrast, smaller stars, like our Sun, consume their hydrogen fuel at a slower and more efficient pace. This allows them to shine steadily for billions of years. Meanwhile, giant stars may burn brightly, but they live fast and die young, often ending their lives in spectacular explosions known as supernovae.

Blue Supergiants:

Due to their enormous masses, stars like blue supergiants have relatively short lifespans, often lasting only a few million years. Their intense gravitational forces and extremely high core temperatures cause them to burn through their nuclear fuel—mainly hydrogen—at a much faster rate than smaller stars.

Red supergiants:

They are massive stars in a late stage of their life cycle, and they are relatively young in astronomical terms—usually no older than about 25 million years. Because these stars are so massive, they evolve quickly compared to smaller stars, moving rapidly through their earlier stages of stellar evolution and entering the red supergiant phase much sooner.

Main sequence stars:

Stars such as our Sun, are capable of remaining stable and active for several billion years.

The main sequence phase is the longest and most stable period in a star's life. For a star like the Sun, this phase lasts for about 10 billion years. Our Sun is currently about 4.6 billion years old, meaning it is nearly halfway through its main sequence lifetime. Stars that are smaller than the Sun can remain in this phase for even longer—sometimes up to tens or even hundreds of billions of years—because they burn their fuel more slowly.

Red dwarfs:

These can live for hundreds of billions to even trillions of years, which is far longer than the current age of the universe, estimated to be about 13.8 billion years. This means that not a single red dwarf has yet completed its full life cycle— we've never seen one reach the end of its life.

The reason red dwarfs live so long lies in their small mass and slow energy consumption. Some of the smallest red dwarfs, especially those with masses around 0.08 to 0.1 times the mass of the Sun, are predicted to continue fusing hydrogen for up to 10 trillion years— almost a thousand times longer than the Sun's entire main sequence lifespan.

When stars reach the end of their life cycles, they often leave behind some of the strangest and most fascinating objects in the universe. The death of a massive star is usually marked by an incredibly powerful and explosive event known as a supernova. This explosion is not only visually spectacular but also crucial to the chemical makeup of the universe.

In earlier discussions, we explored how a star produces energy through nuclear fusion, the process in which hydrogen atoms are fused together under immense pressure and gravity to form heavier elements. This fusion releases a tremendous amount of energy, which counteracts the force of gravity trying to pull the star inward. As long as there is hydrogen to burn, the star remains stable.

However, when a star exhausts its hydrogen fuel, nuclear fusion slows down, and gravity begins to dominate. The delicate balance that kept the star stable is broken. With fusion no longer strong enough to hold the star up, gravity causes the star to collapse inward. This collapse triggers a supernova explosion, violently ejecting the star's outer layers into space. In the process, all the elements created during the star's life—such as carbon, nitrogen, oxygen, and even heavier elements like iron—are scattered across the cosmos.

These expelled elements are the building blocks of everything around us: planets, moons, and even living beings. Without supernovae, many of the elements essential for life would not exist. The core of the star, now left without support, collapses into a highly compact and dense object, which can become a neutron star or, if the original star was massive enough, a black hole.

In regions called star clusters, where many stars of different masses and colours exist, we can observe how various types of stars reach the end of their lives at different times. The most massive stars, typically blue in colour, are the first to disappear. These stars burn their fuel quickly and may die in just a few million years. After the blue stars are gone, the yellow stars—which live much longer, often for billions of years—begin to age.

As yellow stars near the end of their lives, they expand and cool, transforming into red giants. This process usually occurs over the last ten million years of the star's life. In cases where the star is especially massive, it becomes a red supergiant, which is a much larger and more luminous version of a red giant. Sometimes these massive red stars are mistaken for entirely different types of stars due to their size and brightness.

Fig. 9 A Star cluster with varying colors

This gradual change in the colour of stars within a star cluster is extremely helpful to scientists. By studying which types of stars are still present and which have already disappeared, astronomers can estimate the age of the cluster. The sequence in which stars of different masses die out serves as a natural clock, offering valuable insight into the history and evolution of stellar groups.

In the next section, I will go into more detail about the different types of dead stars, such as white dwarfs, neutron stars, and black holes, and explore the tremendous influence these stellar remnants have on the structure and future of the universe.

TYPES OF DEAD STARS

In earlier lessons, we've explored how stars are born, how they live out their lifespans, and how they eventually die. But what happens after a star dies? Does it simply vanish, or does it leave behind something? In fact, stars do leave behind remnants—almost like a kind of cosmic fossil or corpse—but exactly what is left behind depends on the mass of the star and how it ends its life.

The fate of a dying star is governed by very specific physical rules, and one of the most important breakthroughs in understanding this came from the work of an Indian astrophysicist named Subrahmanyan Chandrasekhar. As a young graduate student in the early 20th century, Chandrasekhar made a groundbreaking discovery about the limits of a dying star's ability to resist gravity after it runs out of nuclear fuel.

He asked a crucial question: how massive can a star be and still resist collapsing under its own gravity once fusion has stopped? His detailed calculations showed that if the remnant core of a star—what's left after it sheds its outer layers—is more massive than approximately 1.4 times the mass of our Sun, it can no longer support itself. The internal pressure, even from the tightly packed electrons in a white dwarf, would not be strong enough to hold back the force of gravity.

This critical threshold is now known as the **Chandrasekhar Limit**.

Chandrasekhar Limit:

The Chandrasekhar Limit is named after Subrahmanyan Chandrasekhar, one of the greatest child prodigies in scientific history. Showing remarkable talent from an early age, Chandrasekhar graduated with a degree in physics before turning twenty, a clear sign of his extraordinary intellect. His early achievements earned him a prestigious Government of India scholarship to study at the University of Cambridge in England. In the autumn of

1930, at just nineteen years old, Chandrasekhar embarked on a voyage to England, where, during the ship journey, he completed the bulk of the work that would later earn him the Nobel Prize in Physics.

Fig. 10 Subrahmanyam Chandrasekhar

To understand the importance of Chandrasekhar's discovery, it helps to look back to the 1920s, about a decade before his journey. During this time, astronomers had made a surprising discovery about *Sirius B*, the faint, dead companion star to the bright star Sirius. They found that *Sirius B* had an astoundingly high density—more than a million times denser than the Sun. Such an extreme density suggested that the atoms in Sirius B were no longer behaving like normal atoms. Instead, gravitational pressure compressed these atoms so tightly that the star was essentially made of positively charged ions surrounded by a dense "sea" of electrons.

This discovery was first explained by the physicist Ralph Fowler, who would later become Chandrasekhar's graduate supervisor. Fowler showed that the dense star was supported by what is called Electron Degeneracy Pressure, a quantum mechanical effect that prevents electrons from being squeezed into the same energy states. However, Chandrasekhar realized that Fowler had overlooked a crucial factor: at such extreme densities, the electrons inside these dead stars must be traveling at speeds close to the speed of light. This meant that relativistic effects, described by Einstein's theory of relativity, had to be taken into account.

When Chandrasekhar included these relativistic effects in his calculations, he made a spectacular discovery. He found that there was a firm upper limit to the mass of a star's core that could be supported solely by electron degeneracy pressure. This upper boundary is now known as the Chandrasekhar Limit— about 1.4 times the mass of our Sun. If a stellar remnant's mass exceeds this limit, the pressure from electrons is no longer enough to resist gravity, and the star will begin to collapse further.

This finding fundamentally changed how scientists understood the death of stars and the formation of compact stellar remnants like white dwarfs, neutron stars, and black holes. Chandrasekhar's work laid the foundation for modern astrophysics and earned him the Nobel Prize, while also marking one of the most important scientific breakthroughs of the 20th century.

The types of star corpses this study found out are:

White dwarfs:

These are the dense, compact remnants left behind after the death of stars that are below the Chandrasekhar Limit— which means stars with a mass less than about 1.4 times the mass of the Sun. These remnants typically come from medium-sized stars, including stars similar to our Sun and red giants.

As these stars near the end of their lives, their outer layers begin to expand dramatically, causing the star's radius to swell to about 200 times larger than its original size. This phase is often known as the red giant phase, where the star's outer surface cools and glows red due to the expansion.

After this expansion, the star experiences a small but powerful burst— a kind of stellar "breath"—which blows away much of the star's outer material into space, creating beautiful nebulae. What remains after this shedding of outer layers is the star's core, which starts to contract under its own gravity.

Fig. 11 Sirius A (White Dwarf) and Sirius B in a binary partnership

Eventually, this core stops shrinking and settles into a stable, very dense state called a white dwarf. Despite having a radius only about the size of Earth—a few thousand miles across—white dwarfs are incredibly dense, with matter packed so tightly that a single cubic inch can weigh thousands of tons.

The white dwarf is held up against further collapse by a special force known as electron degeneracy pressure. This pressure arises because electrons, squeezed so tightly together, resist being compressed further, effectively supporting the star against the immense crushing force of gravity.

Over time, the white dwarf slowly loses energy and cools down, eventually fading from a glowing state into a cold, dark object. What remains is essentially a hard core made mostly of carbon and oxygen, a remnant that no longer undergoes nuclear fusion and is slowly cooling over billions of years.

Neutron stars:

Neutron stars are the incredibly dense remnants of stars that were slightly more massive than the Chandrasekhar Limit, but have collapsed into an object much smaller and denser than a white dwarf. These stellar corpses are extremely compact, with a typical diameter of only about 10 to 15 miles—roughly the size of a small city—yet they contain more mass than our Sun.

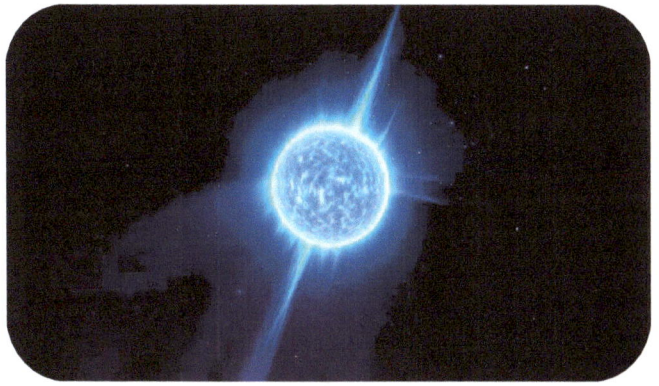

Fig. 12 A Neutron Star

These are formed after the catastrophic explosion of a massive star in an event known as a supernova. When a giant star runs out of nuclear fuel, its core collapses under gravity, triggering a supernova that blows off the star's outer layers. What remains is the neutron star, often covered by a thin but extremely dense outer layer made mostly of iron and carbon.

Beneath this crust lies an unimaginably dense "sea" of neutrons packed tightly together. To give an idea of their density, just a teaspoon of neutron star material would weigh around a billion tons—far heavier than anything we experience on Earth. Neutron stars typically have masses between 1.4 to about 2.5 times the mass of the Sun, although some neutron stars formed from metal-rich stars can be even more massive.

One of the most extraordinary features of neutron stars is their incredibly strong magnetic fields. These fields can range from 100 million to a quadrillion times stronger than Earth's magnetic field. Such intense magnetism can cause dramatic effects, including beams of radiation that sweep through space as the star spins, making some neutron stars detectable as pulsars.

The gravitational field on the surface of a neutron star is roughly 200 billion times stronger than Earth's gravity, making their gravitational pull much stronger and more dangerous than that of white dwarfs. This intense gravity means that neutron stars are among the smallest and densest stellar objects currently known—second only to black holes.

Due to their extreme density, strong magnetic fields, and intense gravity, neutron stars provide unique natural laboratories for studying physics under conditions impossible to recreate on Earth, helping scientists understand the fundamental properties of matter and gravity.

Magnetars:

These are a special and extremely powerful type of neutron star, known for having the strongest magnetic fields in the universe. These magnetic fields are so intense that they can be a thousand trillion times stronger than Earth's magnetic field, making magnetars some of the most dangerous and energetic objects in space.

Fig. 13 A Magnetar

One of the most astonishing features of magnetars is their incredibly fast rotational speed—they can spin around their axis as fast as 1,000 rotations per second, which is about 60,000 revolutions per minute (RPM). This rapid spinning, combined with their powerful magnetic fields, makes magnetars unique among neutron stars.

A dramatic example of a magnetar's power was witnessed in 2004 when, while much of America was asleep, a sudden and intense shockwave hit Earth. This mysterious burst caused about one-tenth of our atmosphere to disappear for nearly a minute. Scientists investigated and eventually traced the origin of this shockwave to a monster magnetar located roughly 50,000 light-years away in our own Milky Way galaxy.

The event was caused by a catastrophic starquake on the magnetar's surface. Due to the enormous stress on the star's magnetic field, a crack formed, causing the upper crust of the magnetar to shift by about one centimetre. Although this might seem tiny, it released an unimaginable amount of energy— equivalent to about 20,000 trillion atomic bombs exploding in just one minute.

If there is a theoretical upper limit to magnetic power anywhere in the universe, magnetars represent that upper boundary. Their extreme magnetic fields affect everything around them, from distorting atoms to creating intense bursts of X-rays and gamma rays that can be detected across vast distances in space.

Magnetars remind us just how violent and powerful the universe can be, and studying them helps scientists better understand the behaviour of matter and energy under some of the most extreme conditions imaginable.

Pulsars:

These are a fascinating type of highly magnetized, rotating neutron stars. What makes pulsars unique is that they emit intense, focused beams of electromagnetic radiation—including radio waves, X-rays, and gamma rays—from their magnetic poles. Because these beams are tightly concentrated and sweep around like the beams of a lighthouse, we can only observe a pulsar's radiation if one of its beams happens to point directly toward Earth as the star spins. This "pulsing" effect is how pulsars got their name.

Inside a pulsar, the matter is incredibly dense—much denser than a typical neutron star—with about a trillion tons packed into just one cubic centimetre. This extreme density, combined with powerful magnetic fields and rapid rotation, creates some of the most energetic and precise cosmic clocks in the universe.

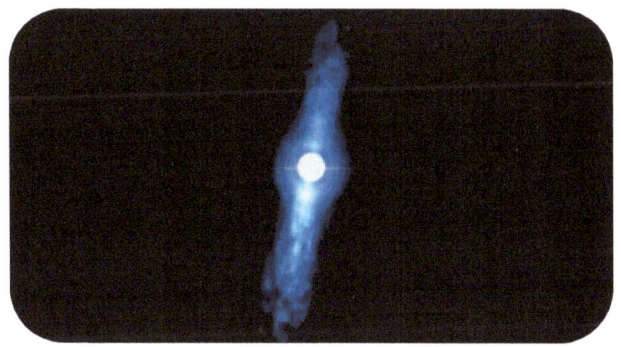

Fig. 14 A Pulsar

One remarkable example was discovered in the galaxy NGC 268420, where scientists found a pulsar spinning at 33 rotations per second (RPS) while locked in a gravitational "binary" relationship with a companion star. The pulsar's intense radiation and particle winds from its poles were so powerful that they heated and distorted the shape of the nearby star, gradually stripping away its material. This process effectively means the pulsar was "killing" its companion star by pulling matter off it.

Because of this destructive behaviour, scientists have given these kinds of pulsars dramatic nicknames like "Black

Widow Pulsars" or "Zombie Pulsars", reflecting how they slowly consume their companion stars.

The fastest known pulsar spins at an astonishing 45,000 revolutions per minute (RPM), which is faster than most man-made machines and showcases the incredible physics at work in these dense stellar remnants.

Pulsars are not just cosmic curiosities; their predictable pulses are used by astronomers for everything from testing the laws of physics to helping navigate spacecraft.

Black holes:

These are perhaps the most famous and mysterious objects among the remnants of dead stars. They form when a star's mass is far greater than the Chandrasekhar Limit, typically several times the mass of our Sun. When such a massive star reaches the end of its life, its core collapses under gravity with an incredible force. If the remaining core is about two-thirds or more of the star's total mass, it can no longer resist its own gravitational pull, causing it to collapse

Fig. 15 A Black Hole warping light around it

into a point of infinite density known as a singularity.

At this singularity, the known laws of physics—particularly Einstein's equations of general relativity—break down, meaning we cannot predict what happens inside this infinitely small and dense point. The gravitational pull becomes so extreme that space and time themselves are warped beyond normal understanding.

As the star's core shrinks to a certain critical size, called the Schwarzschild radius or event horizon radius, the gravitational field becomes incredibly strong. At this point, the curvature of spacetime causes what are called light cones—which represent possible paths light can take—to bend inward so drastically that not even light can escape from the region inside this boundary. Because, according to Einstein's Theory of Relativity, nothing can travel faster than light, if light cannot escape the black hole, then nothing else—no matter or information—can escape either.

This boundary around a black hole is known as the event horizon, often described as the "point of no return". Once anything crosses the event horizon, it is inevitably pulled inward toward the singularity, disappearing from the rest of the universe forever.

The term "black hole" was first coined in 1969 by the physicist John Archibald Wheeler, capturing the imagination of scientists and the public alike with the idea of a region in space where gravity is so strong that it swallows everything.

Black holes come in different sizes—from stellar black holes formed by single massive stars, to supermassive black holes found at the centres of galaxies, including our own Milky Way, which can have masses millions or billions of times that of the Sun.

STORY OF ELEMENTS

W hen you think of your car, you probably don't imagine that it's made of elements that were once part of stars. Yet, every single material used to build it has a cosmic origin. Let's break it down:

- Iron and Carbon for the strong steel chassis.
- Copper for all the essential wiring.
- Gold for the precise electronics and chipsets.
- Lead for the sturdy battery.
- Iridium for the durable spark plugs.

These are just some of the elements that make your car run. But where do these elements come from? Surely, they must have all been created in stars, as I've mentioned before, right? Well, it's more complicated than that. Let me explain.

Long ago, during the universe's Dark Ages, no stars existed—only hydrogen, the simplest and lightest element. Then, the first stars, known as Population III stars, began to form. These early stars were huge, much larger than our Sun, and burned through hydrogen quickly. When they exhausted their fuel, they started creating heavier elements like helium, carbon, nitrogen, and oxygen through a process called nuclear fusion. This fusion process could only go up to iron, which has an atomic number of 26. Unlike lighter elements, fusing iron doesn't release energy; instead, it consumes energy. This makes iron a kind of "poison" for stars.

When a star has a significant amount of iron in its core, it can no longer produce the energy needed to counteract its own gravity. The star collapses in on itself and explodes as a supernova. This explosion scatters the elements the star produced over its lifetime into space.

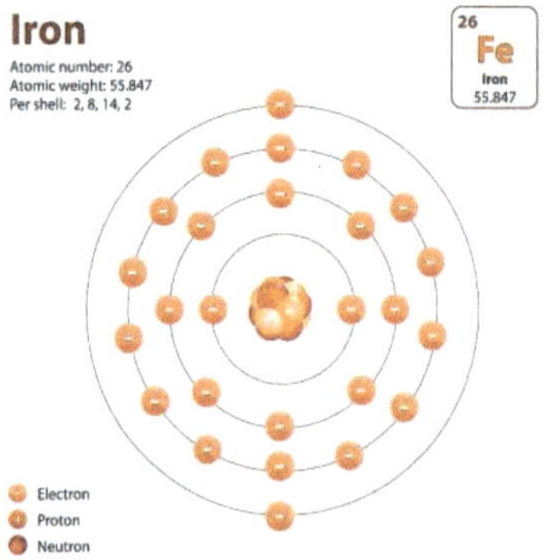

Iron
Atomic number: 26
Atomic weight: 55.847
Per shell: 2, 8, 14, 2

26
Fe
Iron
55.847

- Electron
- Proton
- Neutron

Fig. 16 An Iron Atom

These elements, including iron, carbon, and oxygen, become part of future stars, planets, and eventually—your car. So, the iron and carbon in your car's steel chassis were formed in stars and blasted into space by supernovae billions of years ago.

But what about copper, the metal used for wiring? Copper is even heavier than iron, so how is it made? The early stars I mentioned didn't create elements heavier than iron because it requires an enormous amount of energy to make them. But after the first stars exploded, the elements they made were ejected into space, mixing with gas and dust clouds that would later form medium-sized stars like our Sun. During most of a star's life, the heavier elements, like carbon and iron, stay untouched. However, in the last 10 million years of a star's life, it expands into a Red Giant, and conditions become favourable for forming heavier elements near the surface. This is where copper comes into play.

When a star runs out of lighter elements to fuse, it starts using neutrons—particles with no electric charge—to stick to iron atoms and increase their atomic number. Neutrons can penetrate an atom's nucleus without being repelled, unlike protons, which are positively charged. Once a neutron is attached, it undergoes neutron decay, where it loses its negative charge and transforms into a proton, creating heavier elements like copper. When the star finally explodes in a supernova, the copper is released into space.

For gold, the process is even more fascinating. Gold is much heavier than iron and copper, and it turns out that stars alone can't create it. According to NASA astronomer Jonah Berger, gold and other heavy elements like iridium and lead are produced in the universe's most violent events—neutron star mergers. Since neutron stars are the remnants of massive stars that have exploded in supernovae, having ultra-dense cores, when two of these neutron stars collide, the resulting explosion is so powerful that it can create elements like gold in massive quantities. In fact, when NASA observed a neutron star merger, they found that the explosion produced more gold than the mass of Earth! These heavy elements are scattered throughout space, eventually becoming part of planets, asteroids, and yes—even the gold plating in the electronics of your car.

Many cars today don't just run on gasoline; some use electricity stored in lithium-ion batteries. But lithium is a rare element, and its creation is different from other elements. You can't simply fuse hydrogen and helium to make lithium. Instead, universe uses the most dangerous type of radiation known to man: cosmic rays, which are high-energy particles traveling at near the speed of light, play a role. When a star bursts in a supernova, the explosion releases a tremendous amount of energy, and some of this energy is transmitted to lone protons, turning them into highly energetic particles. These particles behave like natural high-energy accelerators. When these cosmic rays travel through space and strike objects, including living organisms, they can cause serious damage. If they were to strike human DNA, for instance, they could tear it apart, potentially leading to mutations or other harmful effects due to the high energy they carry. Such power can break apart even atoms, like carbon, into smaller ones. This process, known as spallation, can produce lithium atoms by splitting apart heavier elements.

When you step into your car, every element inside it—iron, carbon, copper, gold, lead, lithium—was formed in some of the most extreme environments in the universe. The steel in your chassis was born in the heart of a star and blasted into space by a supernova. The copper in your wiring was forged in the dying moments of a Red Supergiant star. The gold in your electronics was created when two neutron stars collided with unimaginable force. And the lithium in your battery was crafted by cosmic rays tearing apart atoms at nearly the speed of light.

So, the next time you sit behind the wheel, remember: your car is not just a machine built by human hands. It's a cosmic creation, with each element telling a story billions of years in the making. The stars are responsible for everything that makes it run—and they're a lot more than just twinkling lights in the night sky. Every time you turn the ignition, you're connected to the universe in a way you might never have imagined.

In short, the universe itself has gifted us the raw materials for technology and transportation. So, the next time you drive, take a moment to appreciate the cosmic origins of your car, and perhaps even give a quiet "thank you" to the stars. Without them, none of it would be possible.

KILLER STARS

In 2015, a groundbreaking discovery was made by the ground-based satellite dish antenna known as ASSASN (Automated Sky Surveillance for All Supernovae). It detected an exceptionally bright flash of light directed towards Earth. Initially, astronomers speculated that it might be a nearby supernova, but further investigation revealed that the source was actually a staggering 4 billion light-years away. This extraordinary event was named ASSASSIN 15LH, a fitting title given its destructive potential. If planets had existed in the vicinity of this supernova, they would have faced complete annihilation.

The question arose: how could such an intense flash of light originate from such a vast distance? Astrophysicists postulated that the star's remarkable power was primarily due to its rapid spin. This particular star was rotating at an astonishing speed of 1,000 rotations per second. As it approached the end of its hydrogen-fusing phase, the core collapsed, giving rise to a neutron star. Due to its immense speed, this neutron star transformed into a magnetar before the supernova explosion occurred. The combination of a blue supergiant's supernova with the energy of a magnetar produced what is known as a *Superluminous Supernova*.

The devastation caused by a supernova is largely a function of its light. To illustrate this concept, you can perform a simple at-home experiment. Take a magnifying glass and place it between a piece of paper and the sun. You'll notice that the paper begins to burn; that concentrated beam of sunlight can be up to ten times brighter than the sun itself. However, a typical supernova emits light hundreds of thousands of times brighter than our sun. In contrast, the light from a superluminous supernova is about a million times more concentrated than that.

To put this in perspective, a normal supernova has a kill radius of approximately 30 light-years, meaning if one were to occur within this distance from Earth, our atmosphere would likely be obliterated. In stark contrast, the kill radius of ASSASSIN 15LH extended from 500 to 1,000 light-years. This staggering range means that any planets within a thousand-light-year radius of this star would have faced utter destruction. This is the essence of what we refer to as a "killer" star, capable of devastating entire planetary systems across immense distances.

A Supernova can bring destruction through an overwhelming burst of light, while an *Unnova* could spell doom through a chilling descent into ice. Astronomers are exploring a captivating new theory suggesting that a red supergiant star may not end its life with a cataclysmic explosion as traditionally believed. Instead, under certain conditions, it could collapse directly into a black hole. For this to occur, specific criteria must be met: the star must be incredibly massive and spin at an extremely slow rate.

As a red supergiant approaches the end of its life cycle, it may lack the necessary angular momentum to eject its outer layers and instead may retain most of its matter. If such a star has planets orbiting it, those worlds could meet a silent fate. Imagine a rocky planet with liquid water, orbiting a red supergiant at a distance of billions of kilometres. The light from this distant star gently warms the planet, fostering the development of simple life forms like tubeworms, plankton, and microbes in the warm, shallow pools of water. However, red supergiants have relatively short lifespans of only about 10 million years, leaving just enough time for the emergence of basic life.

As the star reaches its final moments, it will collapse into a black hole, its gravitational pull becoming so intense that even light cannot escape. This process would involve approximately 20,000 trillions of trillion tons of matter collapsing into a singularity. The inhabitants of the rocky world would witness a dramatic change in their night sky: a star that once shone brightly would suddenly vanish, plunging the planet into darkness.

The effects of this cosmic event would not be immediate but rather a slow and agonizing descent into a cold death. As the light from the red supergiant fades, the planet would begin to freeze over. First, the land would succumb to the freezing temperatures, followed by the oceans, gradually transforming into ice.

Life on the surface would struggle to survive. Most organisms would go extinct, yet some resilient tubeworms and microbes living near hydrothermal vents at the ocean's depths would endure a bit longer, as they derive heat from the planet's core. Over time, however, even these geothermal hotspots would fail to sustain life. The core would cool, the geological activity would cease, and the planet's atmosphere would collapse under the weight of the frozen surface, leaving it lifeless and desolate.

In this eerie scenario, the gradual transition from a warm, life-supporting environment to a cold, inhospitable wasteland paints a stark picture of cosmic life and death, reminding us of the fragile nature of existence in the universe.

By the intense light of a supernova or the cold darkness of an unnova, planets may meet their demise, but recent discoveries suggest that stars can face their own forms of destruction as well. Recall the colour change observed in star clusters that I mentioned in the third lesson? Astronomers made a fascinating find when they discovered an old star cluster predominantly filled with red stars, but surprisingly, they also detected the presence of blue stars.

Typically, new stars do not form in mature clusters, which raises an intriguing question: how did these blue stars come to be in such an environment? The only plausible explanation is that these stars were acquiring fresh supplies of hydrogen from an external source. But from where could this hydrogen possibly come?

Upon closer examination, astronomers noted that these blue stars were in binary partnerships with white dwarf stars. This setup provides a compelling explanation for the phenomenon: the blue stars were effectively siphoning off hydrogen gas from their white dwarf companions, much like how a vampire feeds on its victim. Here's how the process unfolds: two stars form a binary system, typically consisting of a medium-sized yellow star and a larger yellow star. As the larger yellow star approaches the end of its life, it expands into a red giant. This expansion brings it within the gravitational pull of the medium yellow star.

Once within this gravitational field, the medium yellow star begins to draw in hydrogen gas from the red giant. As a result, the red giant ultimately collapses into a white dwarf, while the medium yellow star, now nourished by this new hydrogen supply, transforms into a massive blue giant, thriving in its new form. Astronomers have dubbed these intriguing 'vampire' stars "blue stragglers."

However, as with any good horror story, there's often a twist. The white dwarf star, not content with its fate, may attempt to reclaim the hydrogen it lost, akin to a zombie trying to reclaim its former self. In this desperate struggle, the white dwarf may accumulate enough mass to ignite a runaway nuclear reaction, leading to a catastrophic explosion known as a Type Ia supernova, taking the vampire star with it in the process. Isn't that a remarkable turn of events?

Furthermore, did you know that supernovae have played a crucial role in helping scientists scale the universe? As stars are observed from great distances, the intensity of their light diminishes according to the distance from the observer. For instance, if a supernova is located twice as far away as another, its light intensity will reduce to one-quarter of that of the nearer supernova. This relationship allows astronomers to measure cosmic distances and gain a better understanding of the vastness of our universe.

SUN - OUR PROTECTOR

M ost people recognize the Sun as the most important star in our sky, and for good reason—it provides Earth with light and heat, making life possible. But the Sun is more than just an energy source. It is the central anchor of an enormous cosmic system—the Solar System—which is far bigger and more fascinating than it first appears.

Many people assume that the edge of the Solar System lies at the orbit of Neptune, the eighth and farthest known planet from the Sun. However, the Solar System stretches much farther than that. In fact, Neptune's distance from the Sun is less than 10% of the true outer boundary of our Solar System!

Beyond Neptune lies a vast, icy region known as the *Kuiper Belt*. This region extends approximately 8 times farther from the Sun than Neptune's orbit. The Kuiper Belt is a wide ring of icy objects, rocky debris, and dwarf planets—with

Pluto being the most famous among them. While it might seem distant and cold, the Kuiper Belt contains important clues about the early history of our Solar System.

But even the Kuiper Belt is not the end. Far beyond it lies the *Oort Cloud*, a massive spherical shell surrounding the entire Solar System. It is believed to be filled with trillions of icy bodies, comets, and leftover debris from the formation of the Sun and planets over 4.6 billion years ago. The Oort Cloud may stretch nearly halfway to the nearest star, making it the true outer boundary of our Solar System.

Scientists theorise that the famous dinosaur-killing comet, estimated to be about 12 kilometres wide, may have originated in the Oort Cloud before being pulled toward Earth by gravitational forces. This shows that even the most distant parts of our Solar System can have a direct impact on life here on Earth.

So, while the Sun is essential for life, the full scale and structure of the Solar System reveal an awe-inspiring cosmic neighbourhood that continues to surprise and fascinate scientists and astronomers around the world.

Sun also plays a critical role in protecting us from some of the most dangerous threats in the universe. It may sound surprising, but the very star that rises in our sky each morning acts as a cosmic shield in more ways than one.

One of the Sun's lesser-known but powerful phenomena is the solar wind. This is not a wind like the breeze we feel on Earth, but a continuous stream of charged particles—mainly hydrogen and helium nuclei—that are blasted outward from the Sun's upper atmosphere, or corona. These particles are accelerated due to intense heat and magnetic turbulence in the photosphere, the visible surface of the Sun.

The solar wind is powerful enough to strip away the atmosphere of a planet if left unprotected. In fact, scientists believe this is what happened to Mars. Billions of years ago, Mars likely had a thicker atmosphere and possibly even liquid water. But with no strong magnetic field to shield it, the solar wind slowly blew much of that atmosphere into space.

Thankfully, Earth—and some other planets—are protected by magnetic fields, also known as magnetospheres. These magnetic shields act as barriers, deflecting much of the harmful solar wind and preventing it from directly eroding the atmosphere or bombarding the surface.

But the solar wind doesn't stop at the inner planets. It travels outward through the Kuiper Belt, past dwarf planets like Pluto, and eventually reaches the outer edge of the Solar System, where it begins to lose energy. At this distant frontier, it forms a massive bubble-like region called the heliosphere.

The heliosphere is essentially the Sun's extended atmosphere and acts as a protective shield for the entire Solar System. It blocks and deflects cosmic rays—high-energy particles from outside our galaxy that can be incredibly harmful. Cosmic rays can damage spacecraft electronics, mutate DNA, and even kill living cells. Without the heliosphere, Earth and the other planets would be constantly exposed to this dangerous radiation.

Cosmic Rays:

Cosmic rays are super-fast, high-energy particles that come from deep space. They travel through the universe at nearly the speed of light and can come from exploding stars or even from outside our galaxy.

So, the solar wind is like a double-edged sword. It's powerful and potentially destructive, but it also creates one of the most important layers of protection we have in space. Without it, the Solar System would be a far more dangerous place.

Magnetosphere:

A magnetosphere is a bubble of magnetic field that surrounds a planet. Earth's magnetosphere is created by the movement of molten iron deep inside its core. It acts like a giant shield, protecting us from dangerous particles like the solar wind and cosmic rays.

Next time you see the Sun rising in the sky, remember—it's not just giving you another bright day. It's also working silently and constantly to shield you from invisible dangers lurking in the vastness of space.